Adding Up the Distance: Critical Success Factors for Internet-Based Learning in Developmental Mathematics

Rob Foshay
Stella Perez

League for Innovation in the Community College
PLATO Learning, Inc.
2000

ISBN 1-931300-19-4

In the Fall of 1999, eight community colleges implemented an Internet-based mathematics curriculum developed by PLATO Learning, Inc. as part of their developmental studies programs. The programs used a variety of structures, ranging from "pure" distance education to various mixtures of classroom and computer-based work on campus. This study characterizes the components of each program and examines the results of student outcomes and subjective reactions of faculty and learners. The aim of the study was to identify critical success factors for Internet-based developmental studies mathematics programs.

Among institutions of higher education, a defining characteristic of American community colleges is the open door through which any and all adult students may gain access to educational and social opportunities. This door stands ajar as a fundamental mission of the community college. Those of all educational levels, backgrounds, and experiences enter, and those requiring remedial and/or developmental education are rarely turned away.

It is no secret that many students who enroll in college are underprepared for the academic rigor of college work. The public debate about remedial education incorrectly assumes that only students who failed to master high school work are enrolling in developmental education courses. Contrary to this myth, two general words sum up the types of students requiring developmental and/or remedial education: number and need. Overwhelming numbers indicate that almost one-half (44%) of students entering 2-year colleges each year require some form of remediation (National Center for Educational Statistics, quoted in Boylan and Saxon, 1998). The number of students who are underprepared for college level work is amplified by the variety of developmental learners' needs. Community colleges describe developmental student demographics as crossing all ethnic, gender, age, and high school experience boundaries. In a recent article, Ikenberry (1999) describes this spectrum of need by example. He claims that while all students in the following list may require developmental education, their needs and academic plans clearly are not the same: a 40-year old woman returning to college for job development skills after being out of school for 22 years, an 18-year old high school student who goofed off in math class, a student who graduated from an inner-city high school that did not offer advanced algebra classes, and a highly-skilled math student whose native language is not English and has difficulty reading math questions. All may require developmental education, but their needs and academic plans may be vastly different.

Developmental students are not new to academic institutions, and leading developmental education researchers adamantly stress, "The problems associated with developmental or remedial education in open-door colleges will not go away" (Roueche & Roueche, 1999, p. 14). Practitioners and educational leaders agree that America's growing diversity and demands for a highly skilled workforce drive more and more students through the open doors of community colleges. Global economic forces have changed the need for industrial era work skills into a growing demand for computer and technology-based expertise in today's Information Era. These transformations require training and development of all sections of the population and fuel the need for programs, services, and options leading to beneficial remediation and college-level success for students.

Concurrently, a powerful tool impacting all social and organizational functions is advancing on the grounds and through the air of American learning institutions. The Internet holds great potential to change the culture, curriculum, and course of educational institutions around the world. For the developmental studies learner, the flexibility and personalization of distance education could allow instructors to reach learners who do not now enroll in, or succeed in, campus-based developmental studies programs.

At the crossroads of the developmental education sector and the application of technology resources are two partners, the League for Innovation in the Community College (League) and PLATO Learning, Inc. In the spring of 1999, the League and

PLATO® initiated an action research project exploring the questions and challenges of implementing successful distance learning developmental math programs for community colleges across the country.

Project Goals

During the past decade, technology has been used in many student programs for a variety of learning goals. However, development and delivery of distance learning programs in developmental studies is still in its infancy. Although some educators may contend that the needs of developmental learners cannot be met successfully via distance education, the issue remains largely untested. Great debate and uncertainty surrounds the types of developmental learners who might succeed in distance education, or how the structure and supports which are typical of successful campus-based developmental studies programs can be implemented in a distance education context.

The purpose of this project was to explore critical success factors for computer-based distance learning in developmental math programs during a summer trial implementation session and a full fall semester term. Mathematics was chosen because it is the subject area of perhaps widest need in developmental studies, and because its content and measures are relatively well defined. College participants, League research team members, and PLATO® service teams outlined four areas of investigation:

- Development of effective, individualized, open entry/open exit programs for developmental students via distance education
- Cultivation of learners' motivation through the use of technology in developmental studies programs using distance education
- Exploration of successful developmental student profiles using distance learning technology
- Effective combinations of campus-based support service and distance learning delivery systems as models of success for developmental learners

PLATO® on the Internet Courseware

PLATO® on the Internet courseware is a modular, self-paced, computer-based learning system that offers students interactive learning opportunities in mathematics, reading, English, and core work skills with over 2,000 hours of instructional content available, all designed for adults. As a comprehensive academic and applied skills courseware system, PLATO® uses computer-based interactive learning processes for student assessment, prescriptive placement, interactive instruction, and evaluative testing and feedback. Delivered on local area networks and stand-alone CD-ROMs, PLATO® is widely used as a core component of campus-based developmental studies programs and freshman/sophomore level courses in post-secondary institutions. PLATO® on the Internet expands this option to any computer with Internet access and a browser.

Methodology

Action or evaluation science is distinguished from basic research. The purpose of basic research is to generate theory and discover "truth." The purpose of action research is to inform action, enhance decision-making, and apply knowledge to address human and social problems (Patton, 1990). Thus, in action research, methodological tradeoffs are likely to favor the "local truth" of the program under study rather than generalizability across settings. While quasi-experimental designs are sometimes possible in action research, fully controlled experimental designs usually are not feasible or desirable. The structures of experimental control are not practical in real-world programs, nor would they even be desirable, because the experimental treatment would create an artificial condition and thus diminish the ability of the study to inform decisions.

Action research studies involving technology often have the stated goal of assessing the effectiveness of the technology. Often, comparison-group designs are used to track achievement in the technology-using program against a conventional program. However, goals such as this are usually ill-informed, and unproductive. The reason is quite fundamental: neither technology-using nor conventional instructional treatments are defined in a way which would allow them to be reproducible. An extremely broad range of instructional activities, some effective and some not, can be employed in either mode of delivery. Furthermore, achievement clearly is a global measure which is of some use to measure general program effectiveness, but cannot separate out the effects of any particular instructional treatment or delivery mode. For these reasons, media comparison studies have very little value, either as basic research or as action research (Foshay 1997). This kind of design would not have been appropriate for the current study.

Studies of instructional methods often are weakened by short duration. Particularly when technology-based methods are used, the learning curve for simply using the technology is substantial, especially on the part of faculty. Furthermore, there is always the possibility that short-term results are due to novelty effects rather than true long-term effects of instruction. For these reasons, the researchers felt it was important to design the study to include a summer pilot component so the project leaders could master the technology and then monitor achievement over the entire length of the following semester.

Action research also needs to be careful in its use of statistical comparisons. In action research studies, treatments often are not well defined or controlled, so interpretation of comparisons–even if statistically significant–can be very difficult unless descriptive research methods are used to model the treatments. In addition, the n for most studies comparing effectiveness is the number of programs, not the number of learners. It makes no sense to aggregate learners across programs, for example, since the treatments are not comparable. The number of programs is often too small to have enough statistical power to justify the comparison. In situations such as this, with a small number of relatively uncontrolled treatments, qualitative research methods often yield the most useful data.

In the current study, the researchers have taken these methodological considerations into account in two ways. First, the researchers recognized from the outset that the study would be primarily descriptive and would rely primarily on qualitative research methodologies. Second, the researchers realized that for most of the research questions, the relevant comparisons would be limited to the number of participating programs (n=8). Statistical power would be far too low for meaningful inference, and the researchers would have to be cautious in interpretation of the quantitative data of the study. As is true with most action research, direct generalization of findings must be limited; however, educators may discover the findings are transferable to their own institutions or situations. The researchers believe that qualitative comparisons were made possible by coordinating the study of eight programs and have the potential to identify issues typical of those found in other programs. The researchers also believe that the study may be of use to others planning similar programs.

College Participants

Selection of participants from League Alliance member colleges was based on specific commitment criteria. Each participating college was asked to designate two faculty members and commit training and service time to research and program development. The research consortium initially included nine colleges; however, only eight colleges fully implemented PLATO® on the Internet (POI) as part of their developmental mathematics program. The colleges were coded as follows:

- Central Florida Community College, Ocala, FL (1)
- Delta College, University Center, MI (2)
- Kapi'olani Community College, Honolulu, HI (3)
- Kirkwood Community College, Cedar Rapids, IA (4)
- Moraine Valley Community College, Palos Hills, IL (5)
- Miami-Dade Community College, Miami, FL (6)
- Santa Fe Community College, Gainesville, FL (7)
- Sinclair Community College, Dayton, OH (8)

Data Collection Model and Processes

The project goals led the researchers to examine a combination of traditional course evaluation processes (course grades, department tests, and/or standardized state competency tests), college profiles, and student and faculty feedback. Therefore, the research design included quantitative and qualitative analysis. A *Data Collection Checklist* was developed, and the data collection forms and materials were made available through the League's website. During the project, students and faculty completed survey documents online and their responses were captured in a database. A listserv was also established as a primary communication tool among faculty project leaders, PLATO® project consultants, and League research team members. Five primary data sets were collected:

- *College Profile Survey & Narrative*–institutional background, history and profile of mathematics developmental education programs
- *Faculty Participants and Instructor Survey*–opinions of POI content, design, and PLATO® service
- *Student Profile*–learner demographics and educational goals
- *Learner Survey*–faculty feedback and evaluation form
- *Course Outcomes*–confirmation of aggregate success and persistence rates of students enrolled in POI-based courses.

College Profiles

The colleges included considerable diversity of program structure and size. Survey data and profiles of participating programs are summarized in Appendix A. While all participating colleges had well established campus-based developmental math programs, they had varying degrees of history and experience with the use of technology in their developmental studies programs. Some colleges have been actively involved with computer-based applications for over 20 years, while others noted this project as their first attempt to use technology tools with developmental learners. The completed college profiles document an average of 9.5 years experience with computer-based learning.

Faculty training and professional development are widely regarded as critical support functions for technological innovation. Underscoring this point, 78 percent of participating colleges offer more than five professional faculty development opportunities focused on the use of instructional technology, and on average, most offer 16 opportunities per year.

Another relevant support service related to distance learning and computer-based applications is technical assistance. All colleges noted significant technical support for faculty and students involved in the project, though the degree and types of support varied by institution. Distance learning delivery and lab-based hardware/software services were offered through central district functions or college/campus contacts.

No two college implementation or service delivery formats were the same (Table A). Based on organizational culture and student need, each college implemented POI in its developmental math studies in a distinct format. Six colleges used POI as a complete online/Internet solution for developmental math courses ranging from basic numeric operations to the highest levels of pre-algebra, with textbooks or selected handouts as supplemental material.

Table A: POI Implementation by College

College	POI Implementation Summary
Central Florida Community College (1)	• POI used as supplemental lab sessions for pre-algebra courses; textbooks used • Web page developed as course communication tool • Mandatory group orientation for course introduction • Access to POI content available from home or campus lab • Dedicated technical support services for POI courses • State-mandated placement and outcome exam

Table A: POI Implementation by College

College	POI Implementation Summary
Delta College (2)	• POI used as a primary source of instruction for Algebra I courses; textbook used • Open entry/open exit class format • One-to-one mandatory student orientation • Access to POI content available from home or campus lab • All developmental math students offered opportunity to use POI courseware as self-study practice option during scheduled open-label times
Kapi'olani Community College (3)	• POI used as a primary source of instruction for beginning algebra courses; no textbook used • Web page developed as course communication tool • Mandatory group orientation for course introduction • Access to POI content available only from home • Additional handouts developed to support course objectives • Students required to attend two-hour lecture each month for test preparation and progress reports
Kirkwood Community College (4)	• POI used as nontraditional college math prep course aimed at upgrading assessment scores of students for limited enrollment programs • POI curriculum used as primary source of instruction; student assignments, tests, and progress evaluated by POI courseware • One-to-one mandatory student orientation • Access to POI content available only from campus lab • Final course evaluation based on post-test results from college assessment testing
Miami-Dade Community College (5)	• POI used as a primary source of instruction for introduction to pre-algebra; textbook used • Open entry/exit for mini-term session • One-to-one mandatory student orientation • Access to POI content available from home or campus lab • Majority of participating students enrolled in the course after transferring from more semester-bound traditional courses • State mandated placement and outcome exam

Table A: POI Implementation by College

College	POI Implementation Summary
Moraine Valley Community College (6)	• POI used as a primary source of instruction for basic math/fundamentals of arithmetic and introduction to pre-algebra; textbook used • Mandatory group orientation for course introduction • Access to POI content available from home or campus lab • Faculty members maintained formal weekly schedule 3 times a week for one hour and during this time were available to students for lab work and/or progress meetings • Course objectives included specific POI lessons as graded assignments
Santa Fe Community College (7)	• POI used as a supplemental resource for pre-algebra; textbook used • Mandatory group orientation for course introduction • Access to POI content available from home or campus lab • Students used POI courseware in a college lab-based setting as part of a regularly scheduled course • State mandated placement and outcome exam • Access to POI content available only from home • Faculty maintained weekly routine contact with students and an informal policy of 24-hour reply to concerns and issues
Sinclair Community College (8)	• POI used as a primary source of instruction for pre-algebra; textbook used • Web page developed as a course communication tool • Open entry/open exit course format • Mandatory group orientation for course introduction • Access to POI content available only from home • Faculty maintained weekly routine contact with students and an informal policy of 24-hour reply to concerns and issues

It should also be noted that, although two colleges (Central Florida Community College and Sinclair Community College) had previously offered developmental math through distance learning options, six of the eight participating colleges had no experience integrating Internet and/or other distance learning means as course offerings within their developmental programs. For these six colleges this was a first attempt at distance learning capabilities for developmental math courses. Experience with computer-based learning in lab settings is also relevant to distance learning. The completed college profiles document an average of eight years of program development experience with computer-based learning in labs and other Local Area Network (LAN) systems.

Faculty Participants

Two faculty participants were selected as project leaders from each college. Both attended a two-day POI training session in the summer of 1999. With the exception of Central Florida Community College (CFCC), both project leaders were full-time math faculty members. CFCC's project team consisted of one math faculty member and a technical support staff member from the college learning center. Faculty experiences with instructional technology ranged from computer-based learning novice to distance learning expert, with an average of nine years teaching experience with computer applications.

Project Leader Surveys on PLATO®

Of the eight participating colleges, only Central Florida Community College had previously used PLATO® on the Internet software for a distance learning course. Prior to the study, one college faculty group–Delta College–had used other PLATO® software in a computer lab-based setting with students in developmental math studies. Seven participating faculty groups used PLATO® on the Internet for the first time in developmental math studies and as part of this research project.

As part of the data collection process, project leaders were asked to evaluate POI software based on four criteria:

1) Depth and breadth of curriculum content
2) Student interface and design
3) Ease of navigation
4) Instructional experience

Survey questions were categorized under the four established criteria and tabulated by mean (Table B). The full survey questionnaire and response tabulation can be found in Appendix B. In addition to quantitative response, reported faculty comments, recommendations, and experiences were analyzed and summary results are noted with the accompanying narrative.

Table B: Project Leader Survey Responses

Category	Survey Question Numbers	Mean (1-5; 1=strongly disagree 5=strongly agree
Depth and breadth of curriculum content	1-9	4=Agree
Student interface and design	10-14	4=Agree
Ease of navigation	15-22	4=Agree
Instructional experience	23-32	4=Agree

Curriculum Content. All agreed that the quality and style of POI online provided consistent instruction and generally met all developmental math course objectives, allowing faculty to use POI as a primary source of curriculum. Suggestions included additional topic development and/or expanding specific math concepts online, developing more targeted student progress reports, and the ability to customize and more closely correlate lessons with course objectives. Overall, faculty members were satisfied with POI content, curriculum, and courseware.

Student Interface and Design. Faculty responded favorably to the look and feel of POI software. They agreed that the courseware was free of bugs and errors and used enticing color and appropriate graphics in an engaging way for adult learners. Project leaders also affirmed that POI lessons involved students through frequent, meaningful interaction and feedback rather than through passive reading and mouse-clicks.

Ease of Navigation. Barring the start-up technical challenges associated with Internet access and hardware configurations detailed by many project leaders, once online, faculty felt that POI was easy to use, lessons had a consistent style and quality, and the application used standard keystrokes. Respondents noted that, overall, POI was easy to navigate and students rarely seemed confused or trapped by the system.

Instructional Experience and Comments. In addition to Likert-scale survey questions, project leaders were asked to respond to a series of open-ended questions regarding their best teaching experiences with POI. Project leaders were also asked to share the corollary or what they liked least about teaching with POI. The surveys included questions related to student contact, including frequency and mode of communication, and descriptions of their most effective strategies with distance learning students. Finally, project leaders submitted suggestions and changes related to PLATO® courseware, learner management functions, and/or teaching with POI (see Appendix A).

Students

Data gathered about participating students included a profile, a survey of student experiences, and final course standing. Only those students with complete profiles, surveys, and tracked performance were included in the final analysis, for a total of 185 students across eight college programs. Students who dropped the course or withdrew too early to receive a course grade did not have complete data and were not included in the study.

Student Profiles

Student profiles were collected at the beginning of the course semester to capture the demographics, characteristics, and backgrounds of the 185 student participants (Table C). The identification of differences in POI students from their on-campus counterparts in developmental math programs may help develop a profile of the successful distance education developmental learner.

Developmental education learners enter community colleges with rich cultural histories and experiences, and they indeed reflect America's emerging society of diversity. In Table C, these variables are compared to national developmental education learners as an indicator of representativeness of the students in the study. Although statistical comparisons to national or campus profiles are inappropriate given the constraints of the study, the differences from national averages in this study may suggest that the profile of the distance education learner differs from that of the on-campus population. If this is the case, then further study may show that distance education developmental studies learners are older, more often female, and more often high school graduates than the national norms. The researchers suspect the differences in ethnicity observed may reflect access to a computer rather than any cultural bias for or against distance education.

Learner Surveys. In addition to the instruments used to collect general demographic information, a Learner Survey was distributed to student participants. Of the 185 students that participated in this study, 100% completed the Learner Survey.

Experience Level. As part of a more comprehensive analysis of distance learners' choices and needs, researchers asked students to comment on educational and personal experiences. The results of the data are outlined in Table D (p. 18).

Table C: Student Profiles

	POI Project		*National Average	
Gender	Male = 32%	Female = 68%	Male = 45%	Female = 55%
Age	27 years		24 years	
Ethnicity–				
African-American	3%		23%	
Asian	16%		3%	
Hispanic	10%		6%	
Native-American or Alaskan Native	2%		1%	
White	69%		67%	
High School Graduate	89% HS Graduates		74%	

Adapted from "Remedial Education: An Undergraduate Student Profile," by L. Knopp, 1998, *Research Briefs*, 6, p. 3.

Motivation. Motivational factors related to distance learning were also documented through the Learner Survey (Table E, p. 19). Results of the full tabulation can be found in Appendix C. POI student's expectations and measures of satisfaction were evaluated by formal questionnaire with open-ended response topics. The questions were analyzed using seven patterned categories:

- Comfort level with computers, the Internet, and technology resources
- Evaluation of orientation processes and support services
- Alignment of course objectives with POI lesson assignments
- Relationship of personal confidence to distance learning courses
- Where and when POI students complete the majority of their assignments
- Best features of POI
- Least preferred features of POI

Table D: Student Experience

College	POI Student Experiences	
Years of High School Math Completed	Average = 3 years	
Previous Math Courses	Algebra I	78%
	Algebra II	38%
	Geometry	59%
	Trigonometry	17%
Last Attended School	Average = 1 year	
College Experience	57% prior college experience 43% none	
Work Experience	81% worked 19% did not work	
	Weekly Work Schedule	
	less than 10 hours	20%
	10-20 hours	16%
	21-30 hours	16%
	31-40 hours	29%
	more than 40 hours	19%
Children at Home	38% one or more 62% none	
College Career Goals	83% specific career goals 17% undecided	

Course Outcomes

As an indicator of progress and success, final grades from the Fall 1999 semester were collected from all participating colleges. For those programs offering an open entry/open exit format, in-progress measures were tabulated and assigned. Given the range of treatments, types of courses, and the range of outcome measures, it is important to caution against broad generalization of the aggregation of noted results.

Table E: Student Expectations and Satisfaction

Category	POI Student Response Summary
Comfort Level	POI students claimed high levels of comfort with computers, the Internet, and technology resources, and minimal levels of anxiety while participating in the study. Frequent faculty contact through e-mail and helpdesk contact was positively correlated to student's comfort level with distance learning.
Orientation and Support Services	Those who attended mandatory formal orientation more positively noted that questions and issues were answered and expressed greater ease of initial logon than those who had no formal orientation.
Alignment of Course Objectives and POI assignments	Students shared more positive comments related to comprehension, time-on-task, and motivation when they recognized direct correlation of their online assignments with course objectives and exams.

Achievement. Tables F and G (p. 20) outline the total population of participants and project outcomes as of January 1, 2000 in aggregate and by college.

Formative Conclusions

Adding up the Distance began by exploring college administrative processes, instructors, and students as independent variables of distance learning developmental math programs. Analysis of the data allows some tentative conclusions to be drawn about the interdependent relationship of college resources, instructors, and learners in successful distance learning models.

The most general conclusion is that the colleges that fully integrated POI curriculum with existing course objectives in their developmental math programs were the most successful

Table F: Project Success Rate

Total POI Participants (n=)	Success Rate (at close of project)	In-progress	Anticipated Success Rate
185	89 (48%)	27 (14%)	116 (62%)

Table G: Project Success Rate by College

College	POI Participants	Success Rate (at conclusion of project)	In-progress	Instructor Developed Final Exam	State Mandated Outcome Assessment
Central Florida Community College (1)	21	12	0	-	-
Delta College (2)	13	10	0	-	-
Kapi'olani Community College (3)	25	9	0	-	-
Kirkwood Community College (4)	25	9	11	-	-
Miami-Dade Community College (5)	16	1	9	-	-
Moraine Valley Community College (6)	18	10	2	-	-
Santa Fe Community College (7)	39	30	0	-	-
Sinclair Community College (8)	28	17	5	-	-
	N=185	89	27		

in using POI with students. The factors that appeared to be most critical to the success of these programs are clustered within six categories outlined below. Each category is accompanied by documented analysis of the research data and is illustrated with select faculty and student participant quotes where appropriate.

A. Development of effective, individualized, open entry/open exit programs for developmental students via distance education

Beyond the traditional functions of student services and development of course objectives, distance education services and curriculum should be enhanced to include a more comprehensive plan with the following variables:

1. Easy Access to Internet and Easy Navigational Courseware

Although the majority of students who enrolled in the distance learning courses expressed high levels of comfort and expertise with computer-based applications, successful students cited as beneficial courseware that makes logon/logout functions and transition from lesson to lesson as smooth as possible.

Comment: *I think [the software] is well designed. It was easy for me to get my lesson and move through my assignments. The way things are laid out made sense to me.*

2. Technical Support

Again and again, technical support (via college helpdesk or program contact) reigned as the most important factor cited by both students and faculty to program success.

Comment: *The college helpdesk gives me the opportunity to ask and find out questions on my own without having to wait and always ask the instructor.*

3. Alignment of Online Courseware & Course Objectives

The programs that correlated course objectives with POI lessons in a meaningful way, whether as supplemental or primary content, and connected assignments and class activities had more successful outcomes than those programs that used POI as a drill-and-practice exercise.

Comment: *I appreciated the opportunity to do all of the computer tutorials, assignments, and mastery tests related to the course assignments, and ensure that I understand all of the work that I will be tested on.*

4. Individualized Instructional Format

Faculty who used the computer-adaptive components of the POI management system and offered individualized and targeted assignments received higher ratings from learners and more favorable comments on student surveys. The self-paced, individualized, anytime/anyplace functions of distance learning were noted as the "best" features of the project by students and faculty.

Comment: *I can review assignments at my own pace. I enjoy the sense of accomplishment when I complete a section, and I can do all this at my own pace.*

B. Development of successful student profiles using distance learning technology

5. Student Recruitment and Counseling

Proactive selection, preparation, and counseling with students entering distance learning programs were noted as key variables for success and course completion. Student who demonstrated a sense of motivation, time management, and program/academic goals were more successful in the POI project than those who transferred from more traditional courses and wanted to avoid class meetings.

Comment: *I suggest that [the software] be used in a lot more classes! I learned easily from the lessons and felt confident in what I learned. I know what I have to do to pass the class and can work at my own pace to accomplish my goals.*

6. Orientation

Students who attended mandatory group orientations cited fewer technical problems, reported greater ease of navigation, and had more successful program outcomes.

Comment: *Meeting my fellow Internet students and teacher at orientation was important and should be mandatory for classes like this. It reminds you that we're all new at this.*

C. Cultivation of learners' motivation through the use of technology in developmental studies programs using distance education

7. Student Connections

Interactive and frequent contact was concluded as an important condition for success. Although many students appreciated the self-paced and individualized format of the POI program, they were quick to note that when questions or issues were resolved via e-mail or helpdesk, there were higher levels of satisfaction with the course and their expressed comfort level with technology. The more successful POI programs in this study had structured assignment schedules with student contact requirements as part of the calendar of course activities.

Comment: *What I like best about learning from the computer is that you can move at your own pace and if you don't understand the problems you can go back and review the material or you can always e-mail a classmate or an instructor.*

D. Combination of campus-based support service and distance learning delivery systems as models of success for developmental learners

8. Faculty Development

As noted, faculty participants had varying levels of experience with technology and computer-based applications. The colleges that offered more than five professional development opportunities and had faculty who were active in attending workshops and conferences created more successful programs in this project.

9. High Standards of Quality and Content Development

As might be expected, faculty who had experience with distance learning had successful program outcomes; however, in a few instances, faculty who were using distance learning as a developmental math option for the first time were also very successful. From the research data gathered it is concluded that those first-time successful faculty showed great interest in computer-based applications and self-initiated the learning curve of teaching with technology. Rather than tag on a few POI lessons with existing course assignments, they reviewed POI content closely and were actively involved in new curriculum development and content upgrade for their courses. They were also very active in seeking technical support and assistance from the PLATO® helpdesk and their assigned PLATO® educational consultant.

Comment: *One of the best features of the project was the innovation and being on the cutting edge. [The project software could] be used as an instructional as well as assessment tool, and I was able to correspond more closely than ever with my individual students by phone, e-mail, and office visits.*

10. College Leadership & Program Support

Participating colleges that designated priority, support, and commitment of resources for technical investments to this project clearly saw successful responses from both faculty and students. Although transparent to the learners, administrative support was recognized in this analysis of clearing the way for successful implementation, program development, and student access leading to high quality services and learning opportunities for students.

Comment: *I like that my college offers this type of program for us. It makes me believe they know how busy adult students are and how much we need this type of learning opportunity.*

Recommendations for Further Research

That we shall be better and braver and less helpless if we inquire than if we indulge in the idle fancy that there was no knowing and no use seeking to know what we do not know.

Plato

Behind these ten critical success factors is the hard work, dedication to innovation, and commitment to learning shared by administrative, faculty, and student participants. Although the project traces the ideas, progress, and outcomes of students over two short semesters, it also demonstrates that future studies should include larger sample sizes and longer durations of measurement to allow quantitative statistical comparisons between programs. Furthermore, the study shows the critical importance of factors other than computer use or the distance context as such.

Future studies should support larger scale comparisons of the critical success variables emphasized in this study. Finally, it is important for future research efforts in distance learning to examine developmental education, since the special needs and program characteristics of this important group of students should be part of the investment in our college, community, and country's future. If community colleges are to journey from the place-bound world of classrooms and computer labs within campus walls to the anytime/anyplace opportunities of distance learning, it is imperative that studies like this one chart the course and guide actions toward a destination of knowing.

Appendix A: College Profile Survey Summary

Question	College/Summary Response	
How long has your college been actively involved in computer-based learning?	(1)	9 (Years)
	(2)	1
	(3)	20
	(4)	5
	(5)	1
	(6)	15
	(7)	5
Average = 9.5 years	(8)	20
List two of the first developmental math department computer-based learning projects at your campus/ college and the year they occurred.	(1)	PLATO® on the Internet (1992) & Academic Systems (2000)
	(2)	PLATO® (1994) & PLATO® (1995)
	(3)	PLATO® (1979) & Comprehensive Competency Program (1987)
	(4)	PLATO® (1999)
	(5)	PLATO® (1999)
	(6)	Tutorial Software (1989) CBT Practice Tests (1994)
	(7)	PLATO® (1999)
	(8)	Center for Interactive Learning (CIL) (1989)
Does your college offer faculty more than five professional development opportunities per year? If yes, how many?	(1)	(Y)–20
	(2)	(Y)–20
	(3)	(Y)–10
	(4)	(N)
	(5)	(N)
	(6)	(Y)–20
	(7)	(Y)–15
Average = 16	(8)	(Y)-12

Question	College	1997--%	1998--%
How many new students have enrolled in Developmental Math (DM) programs over the last two years?	(1)	43	43
	(2)	68	68
	(3)	46	48
	(4)	42	44
	(5)	36	36
	(6)	25	27
Overall Average 1997 = 47%	(7)	61	68
Overall Average 1998 = 49%	(8)	51	52

Appendix A: College Profile Survey Summary

Question	College/Summary Response		
What is the average success rate of students enrolled in DM programs over the past two years?		1997--%	1998--%
	(1)	58	63
	(2)	59	59
	(3)	47	49
	(4)	50	50
	(5)	55	54
	(6)	49	46
	(7)	65	65
	(8)	62	63
What is the average persistence rate of students enrolled in DM programs over the past two years?		1997--%	1998--%
	(1)	92	91
	(2)	92	92
	(3)	53	54
	(4)	52	55
	(5)	56	59
	(6)	69	67
Overall Average 1997 = 70%	(7)	80	80
Overall Average 1998 = 71%	(8)	73	73
Are the majority of DM courses taught by full-time faculty? If no, what is ratio of adjunct to full-time faculty?	(1)	(N)–36	
	(2)	(N)–66	
	(3)	(Y)	
	(4)	(N)–50	
	(5)	(N)–50	
	(6)	(Y)	
	(7)	(N)–50	
	(8)	(N)-100	
Was student recruitment or filling POI courses ever a challenge?	(1)	(N)	
	(2)	(Y)	
	(3)	(Y)	
	(4)	(N)	
	(5)	(Y)	
	(6)	(Y)	
	(7)	(N)	
	(8)	(N)	
Are you requiring an orientation process for students using POI?	(1)	(Y)	
	(2)	(Y)	
	(3)	(Y)	
	(4)	(Y)	
	(5)	(Y)	
	(6)	(Y)	
	(7)	(Y)	
	(8)	(Y)	

Appendix A: College Profile Survey Summary

Question	College/Summary Response	
Outline the technical support components for students enrolled in your POI course.	(1)	Dedicated technical support program person
	(2)	College–IT support services
	(3)	Campus–IT support services
	(4)	College–IT support services
	(5)	College–IT support services
	(6)	College–IT support services
	(7)	College–IT support services
	(8)	College–IT support services
Assessment and placement process for students enrolled in POI	(1)	(Y)–State Assessment
	(2)	(Y)–College-based
	(3)	(Y)–College-based
	(4)	(Y)–College-based
	(5)	(Y)–College-based
	(6)	(Y)–State Assessment
	(7)	(Y)–State Assessment
	(8)	(Y)–College-based
What types of learner support are embedded in your POI course?	(1)	Online tutoring and course bulletin board
	(2)	Course bulletin board, phone support for students from instructor & PLATO® coordinator
	(3)	Computer tutorials with textbook and videos
	(4)	None
	(5)	Computer tutorials for drill and practice
	(6)	Computer tutorials
	(7)	None
	(8)	Video taped lectures, computer tutorials

Appendix B: Instructor Survey Summary

Question	Summary Response (Based on Average) 5=Strongly Agree; 4=Agree; 3=Neither Agree nor Disagree; 2=Disagree; 1=Strongly Disagree
1. The PLATO course content included what my students need to learn about the topics taught.	4
2. The PLATO courses correlated to my course assignments, textbook assignments, and learning objectives for the course.	4
3. The PLATO course content corresponded to the standard end-of-course test we use.	4
4. Content seemed generally free of errors and inaccuracies.	4
5. Content was generally up-to-date.	4
6. Quality and style of instruction was consistent throughout the curriculum.	4
7. Students generally understood the software instructions.	4
8. There was adequate depth in exercises and tests.	3
9. I used PLATO lessons to reinforce content and math problems presented in my lecture notes and textbook assignments.	3
10. Tutorials involved the students through frequent interaction and feedback rather than just passive reading and mouse-clicks.	4
11. The software was generally free of bugs and errors.	3
12. The courseware used consistent keystrokes and display style.	4
13. Color was used appropriately.	4
14. Graphics were used appropriately.	4
15. Formal orientation programs made a positive difference for students using PLATO on the Internet.	4

Appendix B: Instructor Survey Summary

Question	Summary Response (Based on Average) 5=Strongly Agree; 4=Agree; 3=Neither Agree nor Disagree; 2=Disagree; 1=Strongly Disagree
16. I was able to use student progress reports to identify students who need my attention.	4
17. We developed a full orientation for students enrolled in the PLATO distance learning course.	4
18. I was able to make appropriate individual student assignments on the system.	4
19. Sharing class information and student progress through email, web pages, or other means influenced the success of students using PLATO on the Internet.	4
20. My students rarely seemed confused or "trapped" by the system.	4
21. My students responded well to the PLATO system.	4
22. I found working with the computer was generally a productive, rather than frustrating, experience.	4
23. I enjoyed working with the PLATO computer system.	4
24. The PLATO system played a useful role in my teaching.	4
25. I was adequately trained to operate the PLATO system.	4
26. I would like more training on how to use PLATO to its best advantage in my teaching.	4

Appendix B: Instructor Survey Summary

Question	Summary Response (Based on Average) 5=Before or after each computer session. 4=Before or after most computer sessions. 3=Occasionally, before or after a new unit lesson. 2=At the beginning of each semester or marking period. 1=Maybe one time during the year 0=Never
27. Articulated to the students in some way those prerequisite skills, knowledge, or attitudes needed to fully succeed with their newly assigned PLATO modules.	2
28. Helped the students relate what they were about to learn in their PLATO assignments to their own personal previous experiences.	2
29. Described to the students the specific objectives they were going to learn within their assigned PLATO courses or modules.	2
30. Explained to the students how the skills and knowledge learned within their assigned PLATO modules fit into the overall course lesson goals.	3
31. Clearly identified to the students the rewards and incentives for trying hard and doing well within the PLATO system.	2
32. Explained to the students specific procedures for getting support if they did not understand something they were trying to learn within the PLATO system.	3

Appendix C: Student Survey Summary

Question	Summary Response (Based on Average) 5=Strongly Agree; 4=Agree; 3=Neither Agree nor Disagree; 2=Disagree; 1=Strongly Disagree
1. I am able to sign on to the computer without problems.	4
2. Getting to my lessons is easy.	4
3. The computer is easy to use.	4
4. My questions about the PLATO system were answered during my student orientation.	4
5. The computer lets me do something often (like answer questions) and not just watch.	4
6. I usually can understand what the computer teaches me without help from my instructor.	4
7. The computer gives me help when I need it.	4
8. I can work at my pace on the computer.	5
9. My PLATO lessons relate to my course assignments, textbook assignments, and learning objectives for the course.	4
10. The PLATO system helped me understand math concepts when I couldn't understand my instructor's lecture or my textbook assignments.	4
11. I feel I am studying what I need to on the computer.	4
12. Learning about my class members (through email, Web pages, or other means) helped me succeed in this course.	3
13. Before this course, I felt very comfortable using the Internet.	4
14. After this course, I feel very comfortable using the Internet.	5

Appendix C: Student Survey Summary

Question	Summary Response (Based on Average) 5=Strongly Agree; 4=Agree; 3=Neither Agree nor Disagree; 2=Disagree; 1=Strongly Disagree
15. The computer makes me nervous.	1
16. Working on the computer makes me feel good about myself.	4
17. I would recommend learning from the computer to other students.	4
18. The computer lessons I work with are interesting.	4
19. I try hard to learn from the computer lessons.	4
20. The computer lessons make me feel more confident about doing well in school.	4

References

Boylan, H. R. & Saxon, D. P. (1998). The Origin, Scope, and Outcomes of Developmental Education in the 20th Century. In J. L. Higbee & P. L. Dwinell (Eds.), *Developmental Education: Preparing Successful College Students*. National Resource Center for the First-Year Experience: Students in Transition. Columbia: University of South Carolina.

Foshay, R. W. (1997). *Action Research Analysis of Coordinated PLATO® Programs*. Unpublished manuscript.

Ikenberry, S. (1999). The Truth About Remedial Education. *Community College Journal*, 69 (5), 8.

Knopp, L. (1998). Remedial Education: An Undergraduate Student Profile. *Research Briefs: American Council on Education*, 6 (8), 1 –10.

Patton, M. Q. (1990). *Qualitative Evaluation and Research Methods*. London: Sage.

Roueche, J. E., & Roueche, S. D. (1999). Keeping the Promise: Remedial Education Revisited. *Community College Journal*, 69(5), 12-18.

PLATO Learning, Inc.

PLATO Learning, Inc. creates customized learning solutions for developmental studies, workforce development, adult education, and distance learning programs in community colleges. Hundreds of institutions across the nation are successfully using PLATO instructional technology to advance student achievement and extend education and training services into the workplace and the community.

PLATO Learning has a 37-year heritage of instructional research and development, recently incorporating artificial intelligence principles and sophisticated multimedia and graphics into our services and products. Our computer-based and e-learning products are designed to facilitate the learning process and help adult students reach their fullest academic potential. PLATO educational software offers interactive, individualized instruction in a broad range of subjects, including reading, writing, math, science, and social studies as well as job skills, life skills, and basic technical skills.

To find out more about the PLATO Learning, Inc., visit us at www.plato.com.

The League for Innovation

The League is an international association dedicated to catalyzing the potential of the community college movement. Twenty CEOs from the most influential, resourceful, and dynamic community colleges and districts in the world comprise the League's board of directors and provide strategic direction for its ongoing activities. These community colleges and their leaders are joined by more than 700 institutions that hold membership in the League's Alliance.

The League–with this core of powerful and innovative community colleges and more than 100 corporate partners–serves nationally and internationally as a catalyst, project incubator, and experimental laboratory for community colleges around the world. We host conferences and institutes, develop Web resources, conduct research, produce publications, provide services, and lead projects and initiatives with our member colleges, corporate partners, and other agencies in our continuing efforts to make a positive difference for students and communities. These current programs, along with the League's 32-year history of service to the community college explain why in 1998 *Change* called the League "the most dynamic organization in the community college world."